Each student takes a bag home to bring coins to school.

The students pour their coins in a bin.

Some students bring pennies.

This is a penny.

Dedicated to my students
-Mrs. Dorcely

This book belongs to:

_____

penny

nickel

dime

quarter

The students are excited to see the coin charts Ms. Liz displays in the classroom!

"Children, welcome to
'Bring your Coins to School Month!'", says Ms. Liz.

A penny is worth one cent.

Some students bring nickels.

This is a nickel.

A nickel is worth five cents.

Some students bring dimes.

This is a dime.

A dime is worth ten cents.

Some students bring quarters.

This is a quarter.

A quarter is worth 25 cents.

pennies   nickels   dimes   quarters

The students sort the coins.

The students wear coin headbands.

Where is penny?
Where is penny?
Here I am
Here I am
How much are you worth?
How much are you worth?
Just one cent
Just one cent

Where is nickel ?
Where is nickel?
Here I am
Here I am
How much are you worth?
How much are you worth?
Just five cents
Just five cents

Where is dime?
Where is dime?
Here I am
Here I am
How much are you worth?
How much are you worth?
Just ten cents
Just ten cents

Where is quarter?
Where is quarter?
Here I am
Here I am
How much are you worth?
How much are you worth?
Just twenty-five cents
Just twenty-five cents

The students sing a coin song.

The students play the game, "Roll the Coins".

Ms. Liz asks,
"what can we do with the coins?"

Some students say, "we can go grocery shopping."

At the grocery store, the students see apples.

An apple costs one cent.

The students give the cashier a penny.

At the grocery store, the students see bananas.

A banana costs five cents.

The students give the cashier a nickel.

At the grocery store, the students see carrots.

A carrot costs ten cents.

The students give the cashier a dime.

At the grocery store, the students see oranges.

An orange costs twenty-five cents.

The students give the cashier a quarter.

Shopping is fun!

Get your coins!
It's your turn to go grocery shopping.

Some students say,
"we can use the coins at the laundromat."

**Some students say,
"we can save the coins in a piggy bank."**

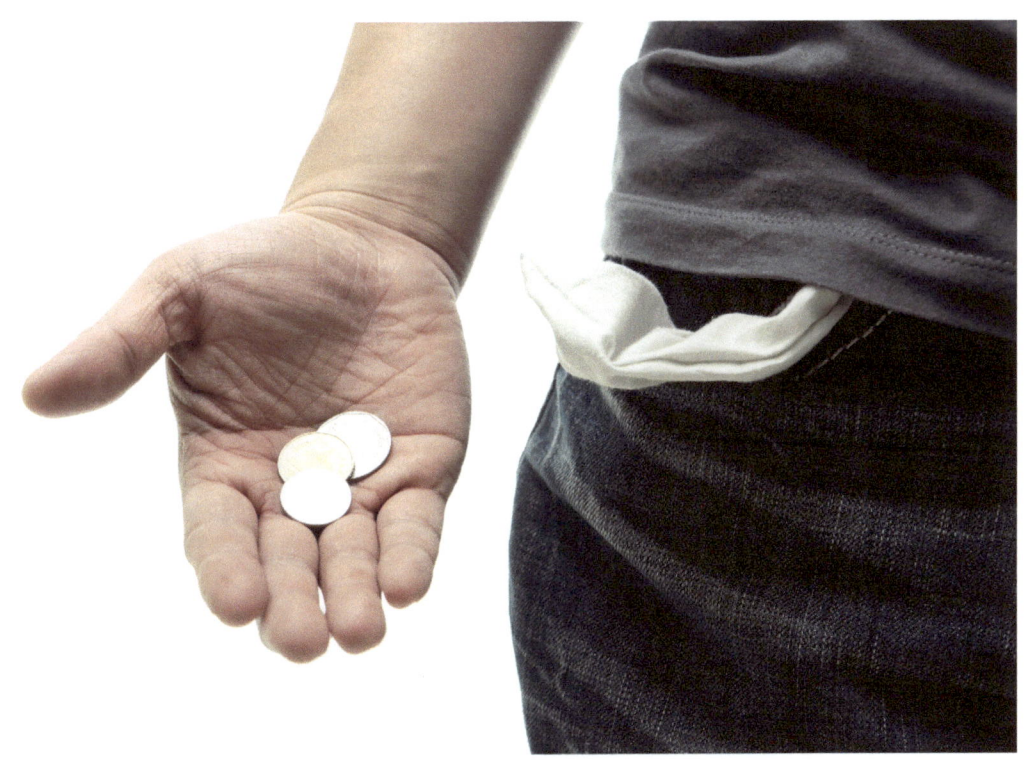

Some students say,
"we can put the coins in our pockets."

**Some students say,
"we can put the coins in the bank."**

The students open bank accounts.

Each student deposits five pennies.

Five pennies are equal to five cents.

**This is one nickel.
A nickel is equal to five cents.**

Each student deposits two nickels.

Two nickels are equal to ten cents.

**This is one dime.
A dime is equal to ten cents.**

Each student deposits five dimes.

Five dimes are equal to fifty cents.

**This is two quarters.
Two quarters are equal to fifty cents.**

Each student deposits four quarters.

Four quarters are equal to one dollar.

This is a one dollar bill.

It is time for you to take your coins to the bank!

Your turn to sort the coins.

My Coins Book

By_____

penny

penny

nickel

nickel

dime

dime

quarter

quarter

1 ¢

5 ¢

10 ¢

25 ¢

 one cent

five cents

ten cents

 twenty-five cents

penny

nickel

dime

quarter

Draw a line to match

Draw a line to match

penny

nickel

dime

quarter

 apple

 banana

orange

 carrot

 **grocery store**

 **coins**

 **laundromat**

 bank

 pocket

 piggy bank

www.ingramcontent.com/pod-product-compliance
Lightning Source LLC
Chambersburg PA
CBHW042026150426
43198CB00002B/74